INTRODUCING

Relativity

Bruce Bassett • Ralph Edney

Edited by Richard Appignanesi

Icon Books UK ◆ Totem Books USA

This edition published in the UK in 2006 by Icon Books Ltd., The Old Dairy, Brook Road, Thriplow, Cambridge SG8 7RG email: info@iconbooks.co.uk www.introducingbooks.com

This edition published in the United States in 2006 by Totem Books Inquiries to Icon Books Ltd., The Old Dairy, Brook Road, Thriplow, Cambridge SG8 7RG, UK

Sold in the UK, Europe, South Africa and Asia by Faber and Faber Ltd., 3 Queen Square, London WC1N 3AU or their agents

Distributed to the trade in the USA by National Book Network Inc., 4720 Boston Way, Lanham, Maryland 20706

Distributed in the UK, Europe, South Africa and Asia by TBS Ltd., Frating Distribution Centre, Colchester Road, Frating Green, Colchester CO7 7DW

Distributed in Canada by Penguin Books Canada, 90 Eglinton Avenue East, Suite 700, Toronto, Ontario M4P 2YE

This edition published in Australia in 2006 by Allen & Unwin Pty. Ltd., PO Box 8500, 83 Alexander Street, Crows Nest, NSW 2065

ISBN-10: 1-84046-757-6
ISBN-13: 978-1840467-57-4

Previously published in the UK and Australia in 2002

Reprinted 2003

Originating editor: Richard Appignanesi

Printed and bound in Singapore by Tien Wah Press

The Conditions of Space and Time

The German philosopher **Immanuel Kant** (1724–1804) delved into the critical limits of knowledge in his revolutionary text, *The Critique of Pure Reason* (1781). He expounded the view that space and time do not exist independently of our consciousness.

Nevertheless, until Einstein, the dominant philosophy of physicists was inherited from **Sir Isaac Newton** (1643–1727).

Newton's Classical Laws of Physics

Newton was arguably the greatest of physicists and mathematicians. He contributed significantly to optics, formulated his three laws of motion, and developed differential and integral calculus independently of **G.W. Leibniz** (1646–1716). But, in terms of understanding Einstein's relativity, Newton's **law of universal gravitation** is the most crucial for us.

BEFORE NEWTON THE MOTION OF THE PLANETS IN THE HEAVENS WAS CONSIDERED A MYSTERIOUS ISSUE DISLOCATED FROM THE EVERYDAY WORLD

Johannes Kepler (1571-1630)

I HAD ALREADY DISCOVERED LAWS FOR THE MOTION OF THE PLANETS...

$$\frac{D^3}{T^2} = K$$

YES, BUT WHAT YOU DISCOVERED WERE EMPIRICAL LAWS WITHOUT THEORETICAL EXPLANATION

A famous but untrue story has Newton sitting under an apple tree when his great discovery of gravity literally hit him on the head.

The specific importance of Newton's law of universal gravitation is that it explains and unites several phenomena within a **single theory**. This quest for a single unifying theory would become the driving force of 20th- and 21st-century physics.

The Law of Gravity

Newton's law of universal gravitation states that the force of gravity (F) between two objects of masses m and M is given by …

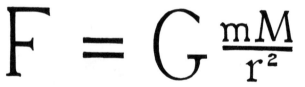

$$F = G\frac{mM}{r^2}$$

where r is the distance between the two objects and G is Newton's constant. G is very small since gravity is very weak.

There are at least two implications to this law of gravity …

Newton took several things for granted in his theory. While the earth was no longer the centre of the universe – and had not been so in the eyes of many scientists since **Nicolaus Copernicus** (1473–1543) – it was assumed that space and time were fundamentally **different** things and that both were **absolute**, set in marble.

HENCE, FOR NEWTON – AND FOR THOSE WHO FOLLOWED HIM – *SPACE* AND *TIME* WERE THE ABSOLUTE AND IMMUTABLE STAGES ON WHICH MATTER IN THE UNIVERSE PLAYED OUT ITS GAMES

The idea of unifying the two, apparently different, concepts of space and time fell to Einstein, as we'll later discuss.

Maxwell's Theory of Electromagnetism

Theoretical physics had made significant progress before Einstein. In particular, **James Clerk Maxwell** (1831–79) had unified magnetism with electricity to give **electromagnetism**.

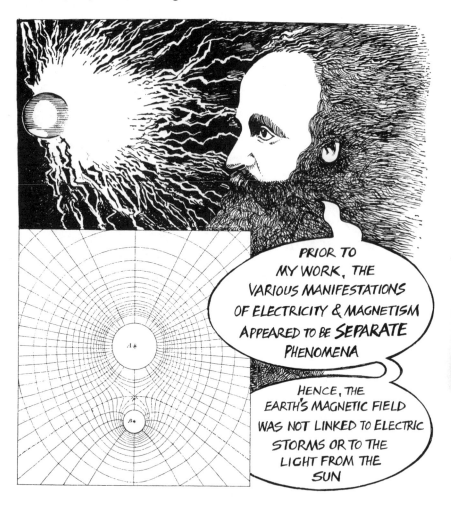

PRIOR TO MY WORK, THE VARIOUS MANIFESTATIONS OF ELECTRICITY & MAGNETISM APPEARED TO BE **SEPARATE** PHENOMENA

HENCE, THE EARTH'S MAGNETIC FIELD WAS NOT LINKED TO ELECTRIC STORMS OR TO THE LIGHT FROM THE SUN

By means of four equations, Maxwell explained all the different manifestations of electricity and magnetism – from the emission of light and electric currents to the earth's magnetic field. Maxwell's equations linked the electric and magnetic fields to each other and showed how each of their various manifestations arose as special cases of a **general theory**.

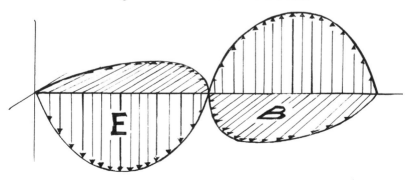

Simple magnetic fields can occur when there is no electric field (and vice versa).

But, in general, if the intensity of an electric field varies in time, it will generate magnetic fields ... and vice versa.

This happens in the case of light, which consists of oscillating electric and magnetic fields propagating through space and time – at the speed of light.

The unification that Maxwell achieved is thus similar conceptually to that of Newton when Newton realized that the force acting on the apple is the same as that holding the earth in orbit around the sun.

Problems in Classical Physics

A number of problems had been identified in this progressive story of physics. One of these concerned gravity itself. Newton's theory of gravity correctly predicted that planets should move in elliptical orbits.

Puzzle of the Atom

The atom was another major thorn in the flesh of physicists. The prevailing picture around the turn of the 20th century was that atoms are made up of a positively charged nucleus surrounded by negatively charged – and much less massive – electrons. The electrons must orbit the nucleus if they are not to fall directly onto the nucleus as a result of the attraction between the opposite charges on the electrons and nucleus.

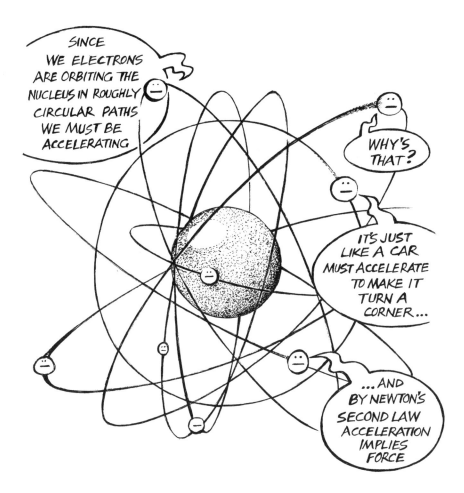

A Major Mystery

Now, from Maxwell's theory of electromagnetism, it was well known that an accelerating charge emits light (or electromagnetic radiation of a different frequency) with an energy that depends on how strong the acceleration is. But, if the electrons lose energy due to the emission of light, then they would begin to spiral inwards and would collapse onto the central nucleus within a thousand-billionth of a second!

The Modern Background

We now have a rough picture of the state of physics in 1905 when **Albert Einstein** (1879–1955) published his account of the Special Theory of Relativity. Einstein did not drop in from a vacuum.

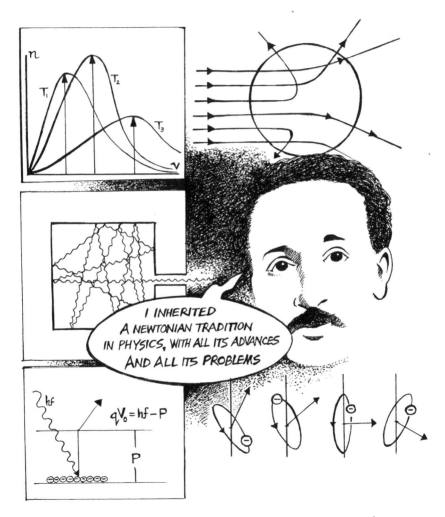

So also, Einstein emerged at a certain juncture in world events, in a particular "climate of mind" that adds some context to his discovery.

Decisive Events

Queen Victoria's death in 1901 signalled the end of a relatively stable period and the beginning of the 20th century's violent releases of energy and accelerated innovations – everything we now call "modern". A dangerous new world arose from two momentous events – first, the "Great War" of 1914 to 1918 …

The second decisive event was the October Revolution of 1917 in Russia which established the Communist Soviet Union there. Communism and the resistance to it from the United States of America and Western Europe set the stage for "Cold War" politics that dominated the world in the second half of the 20th century.

AND FOR ME, THE NUCLEAR BOMBS DROPPED ON HIROSHIMA AND NAGASAKI IN 1945 GAVE HORRIFIC PROOF THAT $E=mc^2$...

A Time of Motion

The energy and disquiet of the early 20th century can be seen reflected in many other headline events. The **Wright** brothers, **Orville** (1871–1948) and **Wilbur** (1867–1912), made their first powered flight in 1903.

Henry Ford (1863–1947) in 1912 brought the assembly-line mass-manufactured Model-T to millions of people.

Pablo Picasso (1881–1973) introduced the revolutionary art of Cubism in 1907, also developed by **Georges Braque** (1882–1963). The British philosophers **Bertrand Russell** (1872–1970) and **A.N. Whitehead** (1861–1947) produced their formidable *Principia Mathematica* in 1910–13, which attempted to re-think mathematics on the rigorous basis of logic.

I ADDED MY OTHER CONTRIBUTION TO PHYSICS IN 1916 WITH MY GENERAL THEORY OF RELATIVITY

We shall outline briefly the essence of special relativity and then focus more on the complexities of general relativity (GR).

Lorentz Transformations

Einstein was a great DIY thinker who made good use of others' often neglected discoveries. The work of **H.A. Lorentz** (1853–1928) is a crucial example.

The remarkable result of special relativity is that it shows how our ordinary intuition about relative motion fails when one is moving near the speed of light, approximately 300,000 km per second. The speed of light is a **fundamental constant** which in Einstein's theory is independent of the speed of the observer.

The Effect of Length Contraction

Alice and Bob are moving at a constant speed, *v*, relative to each other. How does Alice see Bob?

Alice sees the length as shorter than Bob sees it.

Time Dilation

Similarly, we find that the rates at which time flows are different for observers moving relative to each other. But is time slowed down or sped up by moving faster?

However, even to get Bob's time to flow at half the rate of Alice's, he must move at around 86% the speed of light. So it is not an issue for life on earth. However, time dilation has actually been observed, as we'll next encounter.

Observing Muons

Cosmic rays travel from outer space and hit the atmosphere at nearly the speed of light. The collision creates **muons** (strange particles that are like heavy electrons) also travelling near the speed of light.

That's because, when we include the time dilation due to the rapid motion of the muon, we find that the muons live about 20 times longer and hence have enough time to reach sea-level in the observed numbers.

Energy is Mass, Mass is Energy

Einstein's famous relation is that $E = mc^2$. Here m is the "rest mass" – the mass that a body has when it is seen at rest. It shows that mass can be turned into a huge amount of energy. But what if a body is moving rapidly? Then it also has **kinetic** (or movement) energy. In fact, the full form of Einstein's equation is ...

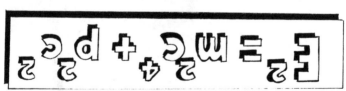

When the particle has no kinetic energy and is stationary, we have $E^2 = m^2c^4$, or more familiarly $E = mc^2$. **Energy is mass and mass is energy.**

But suppose we consider particles which have no rest mass – for example, photons of light. Einstein's formula shows that they still carry energy: $E = pc$, which is the *particle* theory of light. However, light can also be understood in terms of *waves*: $E = hf$, in which h is Planck's constant and f (frequency) $= c/\lambda$, where lambda (λ) is the wavelength of the light, so that we end up with $E = hc/\lambda$ to describe the photon's momentum, p.

The shorter the wavelength, the more momentum the photon has.

This is why skin cancer is caused by ultra-violet (UV) light which has a shorter, or **higher frequency**, wavelength than visible or infra-red light.

But what is Planck's constant, h?

Planck's Constant *h* and Quantum Effects

h = 0.000 000 000 000 000 000 000 000 006 626

Planck's constant, *h*, is a very small number, but it controls the size of *quantum effects*.

One example of quantum scale is that particles behave not only as "particles" (tiny localized lumps of energy) but also as distributed waves (like water waves). This *particle-wave duality* is an effect shared by light, electrons and all other forces of matter.

Another example is barrier penetration …

Quantum and Classical Physics

The speed of light, *c*, and Newton's gravitational constant, *G*, are *classical* in the sense that they have nothing to do with quantum effects. If the speed of light were much smaller, say 10 m/s, then special relativity would have been discovered much earlier. Why? Because everybody would have an intuitive understanding of time dilation and length contradiction.

But if $G = 0$, there would be no more gravitational force at all. No planets or stars would form and the universe would be a very strange place.

Dirac's Idea of Anti-Matter

Let's go back to Einstein's equations which had $E^2 = m^2c^4$. Is there any case when the equation for E^2, rather than E, is important? Yes. **Paul Dirac** (1902–84) noted that when we take the square root to find E, there are mathematically TWO solutions. This is easy to understand: $2 \times 2 = 4$ and $-2 \times -2 = 4$. One can take the negative sign for the square root, giving a NEGATIVE energy. On the basis of this (and a great deal more rigorous analysis), Dirac proposed the idea of ANTI-MATTER with negative energy.

Paul **Dirac**

IN 1932,
MY RADICAL CLAIM,
BASED ON PURE MATHEMATICAL
REASONING, WAS PROVED WITH
THE DISCOVERY OF THE
ANTI-ELECTRON

The Michelson–Morley Experiment

In 1881, **Albert Michelson** (1852–1931) designed an experiment that tested whether the motion of the earth had any effect on the speed of light. In 1887, Michelson and **E.W. Morley** (1838–1923) conducted the experiment at high sensitivity and found that the speed of light did not depend on whether the light travelled with the earth or against the earth's motion.

Constancy of the Speed of Light

One of the most important tenets of special relativity is the postulate that in vacuum the speed of light, c, does NOT depend on the observer. Einstein's postulate replaces Newton's absolute space and time with an absolute speed of light.

The Problem of Simultaneity

As the name itself suggests, relativity implies that there is no unique, absolute splitting of our four-dimensional world into space and time. What does this mean? Well, if time were uniquely defined, we could formulate the idea of **simultaneity** in such a way that everyone would agree with it. Why is there a problem of agreeing on simultaneity?

Slicing Spacetime Differently

Let us imagine that Bob is moving on the line from 1 to 2. Imagine further that Bob passes Alice exactly as the light bulbs go off (as far as Alice is concerned). Then, because Bob is travelling towards point 2, and because the speed of light is the same in all frames, Bob sees the flash of light first from 2 and then later from 1 because he is closer to point 2 when the flash reaches him.

THE TWO FLASHES WEREN'T SIMULTANEOUS FOR ME, DESPITE BEING SIMULTANEOUS FOR HER

WE WILL LATER SEE THIS GEOMETRICALLY AS THE FACT THAT YOU AND ALICE SLICED UP 4-DIMENSIONAL SPACETIME INTO 1-DIMENSIONAL TIME AND 3-DIMENSIONAL SPACE DIFFERENTLY!

The Need for General Relativity

We can now discuss one of the celebrated paradoxes of special relativity that will lead us to the General Theory of Relativity. Consider a pair of twins, one of whom leaves earth in a rocket, while the other remains on earth. The rocket accelerates very close to the speed of light on its way to a star ten light-years away.

A light-year is the distance light travels in a year – a huge distance! Suppose that $v = 0.995c$, so that, according to the time-dilation formula, time travels 10 times **slower** on the rocket than it does on earth.

To the twin on the rocket, the trip to the star and back to earth takes just two years, while for the twin on the earth it seems to take a little over twenty years.

Another Viewpoint

What is the paradox? The twin on earth would have the right to claim that the rocket is **stationary** and it is actually the earth which is moving at nearly the speed of light (together with the solar system). In that case, it should be the twin on earth who sees time tick more slowly, and the twin on the rocket who sees time flow at its normal rate. After all, this is precisely what relativity means!

Out of the Impasse

The twin paradox would seem to lead us into an impasse. There appears to be a symmetry in the problem. The physics looks the same, even if we interchange the twins' arguments, and yet the result for the *amount of time* the rocket twin takes to return changes completely.

A little thought will show the problem. Are the situations and arguments of the twins REALLY interchangeable?

Once the rocket twin is travelling at the CONSTANT velocity of $v = 0.995c$, they are interchangeable.

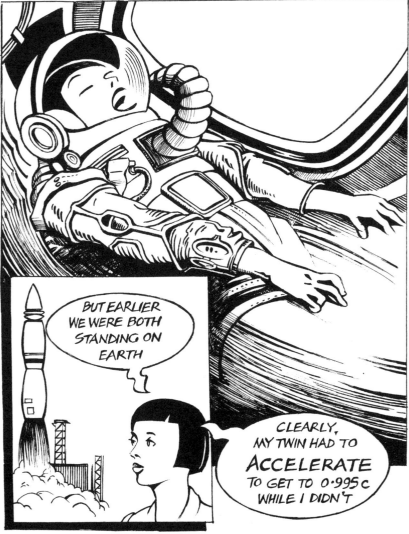

BUT EARLIER WE WERE BOTH STANDING ON EARTH

CLEARLY, MY TWIN HAD TO ACCELERATE TO GET TO 0·995 c WHILE I DIDN'T

Resolving Acceleration

This breaks the symmetry between the twins, showing that you cannot interchange their arguments. We have already made it clear that special relativity does not apply to systems which undergo **acceleration**.

This quest will lead Einstein to general relativity, which he completed in 1916. It is arguably one of the greatest single intellectual contributions to humanity.

The Building-Blocks of General Relativity

We will now consider the basic conceptual building-blocks required to discuss the General Theory of Relativity. Einstein needed ten years to put these together for himself – from 1905 to 1915 – so we will allow ourselves plenty of space to do the same …

Before we begin, a useful philosophy to adopt is that expressed by **John von Neumann** (1903–57) …

This is important to accept when dealing with the strange concepts of relativity. For example, spacetime is four-dimensional – three space, one time. However, there is no way truly to visualize a four-dimensional space, since we are limited to three space dimensions. But there are tricks to aid and give us intuition.

An Infinite Number of Dimensions

But first, what goes through a mathematician's mind when she talks of a space with four, five or even infinite numbers of dimensions? To answer this, let us consider the surface of the earth. The SURFACE itself is two-dimensional – that is, it takes two numbers to specify uniquely where you are on the earth: your latitude and longitude.

BY SAYING YOU ARE AT 40 N 47 AND 73 W 51, I KNOW THAT YOU ARE IN NEW YORK

IF YOU WERE DEEP UNDERGROUND IN A MINE SHAFT NEAR JOHANNESBURG YOU WOULD NEED TO GIVE ME ANOTHER NUMBER ...

Bernhard
Riemann

...YOUR DEPTH BELOW THE SURFACE – BEFORE I COULD PINPOINT YOUR LOCATION PRECISELY

This implies that the earth as a complete SOLID body is three-dimensional, since it takes three numbers to specify any point inside the earth uniquely.

This basic idea can easily be generalized. If you need five points to specify uniquely where you are in a space, then that space is five-dimensional. If you need 25 numbers to specify a point uniquely, the corresponding space is 25-dimensional.

Now, a crucial point to realize is that these spaces need not have ANYTHING to do with the world we live in, and in fact generally don't.

A Thought Experiment

As an aid, let us consider the ancient Greek philosopher **Plato**
(c. 428–347 BC). He suggested that all objects we perceive are
shadows of perfect entities which exist only in our minds.

One can certainly conceive of 35-dimensional spaces, like we did before,
and that space need not have any counterpart in the real world.

To extend this idea, consider the following thought experiment. Imagine that you wish to create a space based on the height of the water under the Rialto bridge in Venice at every moment in time during the 20th century.

This is an ABSTRACT space in that it exists only mathematically, not physically, and is a crucial step in liberating ourselves from the slavery of our own world.

Infinity and Configuration Space

Let us take this further and jump into some of the wonderful complexities of infinity. As we will see later, it seems that cosmological observations MAY favour us living in an infinite universe. In that case, there may be an infinite amount of matter in the universe; an infinite number of atoms.

Slicing Spacetime

However, we have an infinite number of atoms in this example, so the complete space is *infinite dimensional* (4 x infinity = infinity). We need an infinite number of numbers to record uniquely where all the atoms are. This space, though we will not need it, is known as **configuration space** in mechanics, since it gives the configuration of the system.

Notice that by thinking of spaces abstractly, one gives up the need to be able to visualize them in everyday terms.

We can slice up spacetime (which is four-dimensional) into three-dimensional slices which we can visualize.

How to View Spacetime

One of the big advantages of abstraction by letting go of the need to visualize things as they would look in our world is that we can give up the urge to think continually of spaces in terms of them being *inside* bigger spaces.

This is a natural question, from the standard point of view, but not from our new view in which we think of a space as existing completely separately from any other space. Hence, cosmologists usually think of the expansion of the universe only as a *property of spacetime* itself, namely that the distance between any points IN THE SPACETIME is increasing.

Simultaneity is Relative

One of the key ingredients of relativity is that – unlike Newton's view of gravity – space and time are unified into a four-dimensional space which can be sliced, like a loaf of bread, in different ways, to give "space" and "time". But there is NO UNIQUE or preferred way to slice spacetime. This is in fact a *geometrical* way of understanding the lack of simultaneity we observed before.

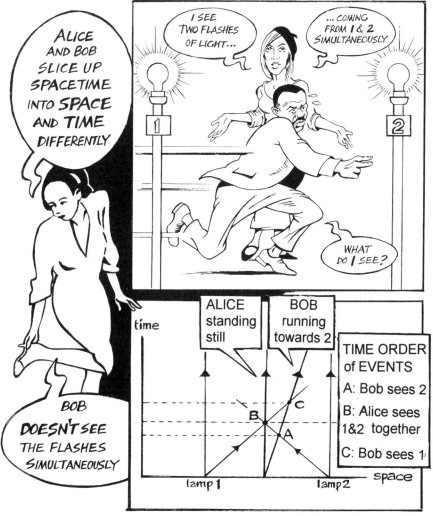

We are now ready to continue following in Einstein's footsteps towards the General Theory of Relativity.

Einstein's Tasks

Between New Year's Eve of 1904 and 1905, Einstein contributed six of the most important papers of the 20th century. Two of these laid the foundations of special relativity (SR). But now he faced the problem of how to extend SR in two directions …

At first sight, these seem very different tasks. But in a brilliant insight, Einstein realized that they are in fact two sides of the same coin. Let's consider the same reasoning that Einstein followed and which he later called "the happiest thought of my life".

Suspending Gravity

If you fall from a window, what do you feel before impact (apart from the rushing of air)? You accelerate towards the ground but feel *weightless*. That is how astronauts train for space – in planes, flying vertically downwards for a couple of minutes.

This led Einstein to suggest that the effects of gravity could be made magically to disappear for short times and for small distances from the observer – in this case you falling with the hammer.

The Equivalence Principle

Let us take this line of thought further. Imagine now that you are blindfolded, lying near the floor of a windowless room which is drifting through space without any forces acting upon it. You are completely weightless. Suddenly you crash to the "floor" and are pinned there.

Perhaps your intuition, like Einstein's, leads you to suspect that you could not tell the difference. These two apparently elementary observations are now known as different aspects of the **equivalence principle**, one of the gems of theoretical physics. With these simple thought experiments, Einstein cut straight to the heart of what needed to be done to enlarge SR to include acceleration and gravity.

Gravitational and Inertial Mass

If you think a little, the equivalence principle means that what we thought were two challenges in extending SR – including accelerating observers and gravity – collapse to the same problem: *an observer cannot tell whether he or she is accelerating due to gravity or another force.*

This has been tested to fantastic precision and should it NOT be true for some magical substance, that substance in "Einstein's room" would solve one's ignorance about whether one were experiencing gravity or being accelerated by a rocket.

Extending Newton's First Law

We have seen that when you fall in a gravitational field you feel weightless, as if there were *no force* acting on you. This led Einstein to the radical idea that gravity isn't a force like other forces! But how do we reconcile this with that most basic of school laws – Newton's first law – which, building on the work of **Galileo** (1564–1642), states that …

The solution to this is so stunningly elegant as to be one of the most beautiful modifications of a theory in the history of physics.

The earth isn't flat and neither is space!

Einstein modified Newton's first law like this …

The problem is that we usually limit ourselves to flat spaces, the legacy of a geometry instituted by **Euclid** (fl. 300 BC). But we all know that the earth is not flat. So why do we limit ourselves by thinking that spacetime should be flat? Well, Newton assumed it, and he was a genius … so it seemed like a reasonable assumption at the time!

In fact, when a space is curved, the curve LYING COMPLETELY IN THAT SPACE of shortest distance between any two points in that space is NOT a straight line. Consider a simple example: the earth.

Examples of great circles are the equator and the lines of longitude. Indeed, there are NO straight lines lying on the surface of the earth!

A Brainteaser

Another example of these ideas is the old brainteaser: what path should an ant take to go quickest from one *inside* corner of a matchbox to the diagonally opposite corner?

A slick technique to solve this is to open up the matchbox and lay it flat.

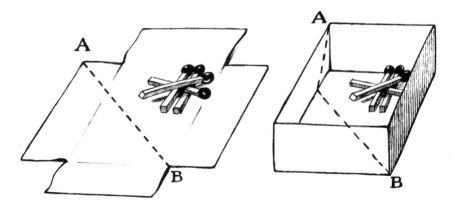

Since it is now a flat space, a straight line DOES give the path of shortest distance which forces the ant to go on a non-intuitive path.

Geodesics

Curves of shortest distance are known in Relativistic jargon as **geodesics**. Hence, all that we need in order to find out how a body will move when there is gravity is to calculate the appropriate geodesic …

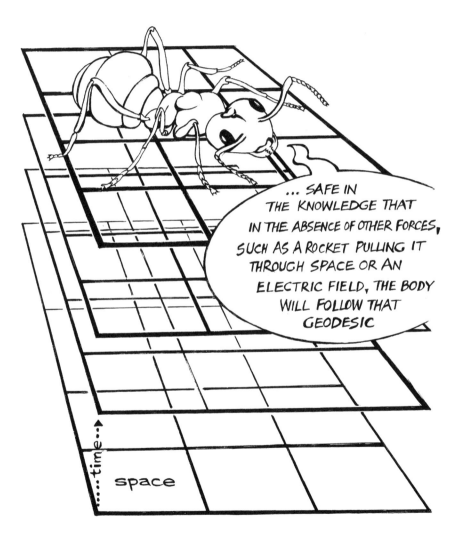

… SAFE IN THE KNOWLEDGE THAT IN THE ABSENCE OF OTHER FORCES, SUCH AS A ROCKET PULLING IT THROUGH SPACE OR AN ELECTRIC FIELD, THE BODY WILL FOLLOW THAT GEODESIC

Spacelike, Null, Timelike

But again we have only part of the puzzle. Where is *time* in our modification of Newton's first law? The path of shortest distance between Mexico City and Oxford could be marked on the surface of the earth for all time (ignoring continental drift). But our modification of Newton's first law talked about bodies moving *in time* along geodesics. Surely this doesn't make any sense!

Well, it turns out that because we have space AND time dimensions, we will need three different types of geodesics.

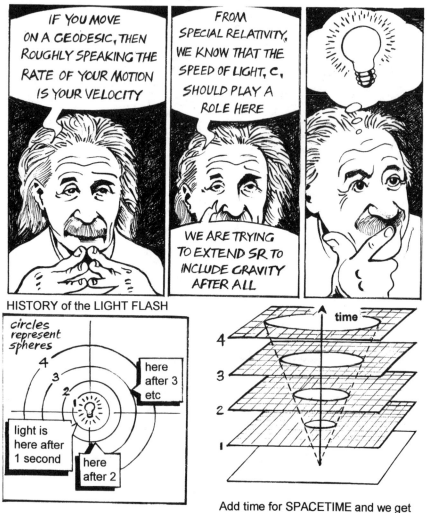

2-D drawing of 3-D space

Add time for SPACETIME and we get the light-cone of the event.

And indeed, the three geodesic classes correspond to motion with velocities less than *c*, equal to *c*, and greater than *c*, respectively known as **timelike, null** and **spacelike** geodesics.

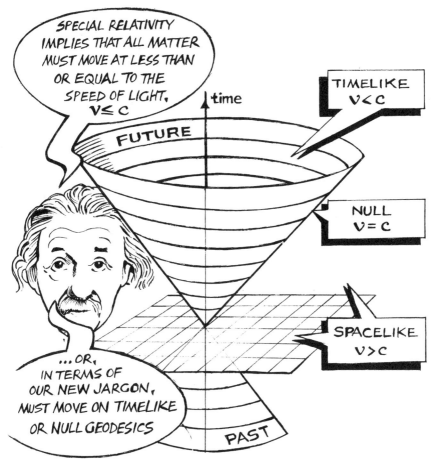

Our great circle from Mexico City to Oxford is a special spacelike geodesic – to travel on that geodesic would require infinite velocity because one is in all places on the curve simultaneously! Our final form of Einstein's modified first law is then ...

All bodies move on timelike or null geodesics unless acted on by a force other than gravity.

This therefore also includes the foundation of SR that *no matter can travel faster than the speed of light.*

Finding the Distance

But geodesics are very difficult to calculate in general. Imagine surveying a complex landscape with hills and valleys, mountains and plains. How is one to calculate the path of shortest distance over this bumpy terrain? Now imagine doing this in four dimensions!

To find the geodesics, we need to introduce a measure of distance. And let us do this by starting with the landscape above.

This is a very familiar distance simply given by Pythagoras' theorem:

$$ds^2 = dx^2 + dy^2$$

where *dx* and *dy* correspond to the differences in *x*- and *y*-coordinates of the two points we are interested in on the map.

Geodesics and the Metric

But remember that geodesics must be defined as curves *lying in the space itself*, not outside it. So, a crow would probably choose to fly over a very deep gorge. Someone trekking across the landscape might choose a path that goes around the gorge and thereby travel a shorter distance than by descending into, and then ascending out of, the gorge.

The mathematical quantity which converts the flat map distances into actual distances on our curved space (here the landscape) is called the METRIC of the space and is unique to that space. We will denote it by "*g*".

The idea of a metric is very common to us. It is a way of converting a universal distance (the distance on a flat space) to distances on a curved space. It is like a taxi meter which converts a fixed amount of time and distance into a cost to the passenger.

Similarly, if you take a taxi in London, it will cost you a whole lot more than a cab in Pune, India, even though you spend the same time and travel the same distance in both cases. The "taxi metric" also depends on your **spatial** position.

Finding the Metric

The same applies to our landscape example. Distances over a very bumpy area will be very different than those on a flat prairie. Indeed, the more bumpy our landscape, the more the true distance differs from our flat map distance. Conversely, the flatter the landscape, the closer the distance is to the universal Pythagorean distance, and the closer the geodesics are to straight lines.

The Metric ...

But what is the metric, g? Well, if we consider our trusty examples of a cylinder and a sphere again we can get some idea. The cylinder is curved in one direction, but not in the lengthwise direction, while the sphere is curved both in the "north-south" and the "east-west" directions. Clearly, if the metric is going to tell us everything about the curvature of a space, it cannot just be a single number at each point of the space, since otherwise how would it tell us the cylinder and sphere are different?

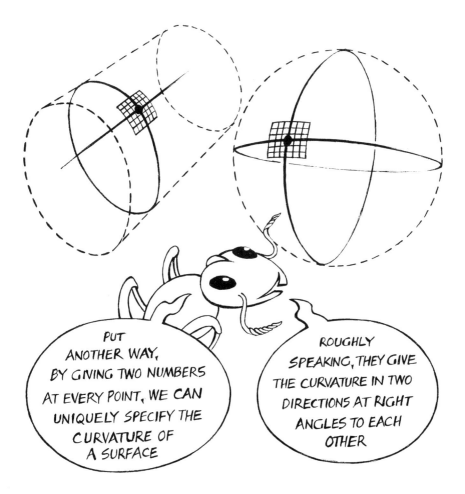

PUT ANOTHER WAY, BY GIVING TWO NUMBERS AT EVERY POINT, WE CAN UNIQUELY SPECIFY THE CURVATURE OF A SURFACE

ROUGHLY SPEAKING, THEY GIVE THE CURVATURE IN TWO DIRECTIONS AT RIGHT ANGLES TO EACH OTHER

Both numbers are part of the metric, a bit like a bicycle has two wheels.

67

The Metric in Four Dimensions

We will designate these two numbers representing the metric by g_{xx} and g_{yy} to show that they are associated with curvature in the x- and y-directions of our co-ordinate system (which is in fact arbitrary). Now, in four dimensions, things are even more complicated, since a four-dimensional space can be curved in four different directions.

SO WE NEED MORE THAN TWO NUMBERS TO UNIQUELY SPECIFY THE CURVATURE AT ANY POINT

IN FACT, WE NEED TEN! SO IF WE THINK OF THE METRIC AGAIN LIKE A BICYCLE, THEN THIS TIME IT HAS TEN WHEELS!

So, if we take small steps *dx* and *dy* in the *x* and *y* directions on our map, we can use the metric to figure out what distance that corresponds to on our curved space by calculating the sum:

$$(ds)^2 = g_{xx}(dx)^2 + g_{yy}(dy)^2$$

So with the metric of the space (or spacetime) known, we can use some advanced techniques to find the geodesics – or we can at least write down the equations obeyed by the geodesics. But, like mystic runes solving these equations is generally extremely difficult and can only be done approximately, using a computer.

Spacetime Geodesics

Until now, our everyday analogies have served us well in our discussion of the geodesics of curved spaces. But, now we must let go and take a deep plunge into the strange world of the geodesics of SPACETIME. It turns out that space and time are not completely equivalent even in relativity.

The strange element that time brings to the discussion of geodesics is that it changes our neat theorem of Pythagoras even in a FLAT SPACETIME.

In space, Pythagoras states that $ds^2 = dx^2 + dy^2 + dz^2$ (in three space dimensions). What happens if we want the distance between two events (t,x,y,z) and (t',x',y',z') in spacetime?

With Newton and **Carl Friedrich Gauss** (1777–1855), **Georg (Friedrich Bernhard) Riemann** (1826–66) has a strong claim to being the leading mathematician of all time. After Fermat's last theorem was solved, the Riemann Hypothesis, which concerns the properties of prime numbers, became the biggest unsolved conjecture in mathematics. The Clay Foundation offers a million-dollar prize for proving it true (and nothing for proving it false).

Including Time

Riemann helped to develop much of the geometric techniques used by Einstein in formulating General Relativity. In Riemannian geometries, the distance between two points does not have to be positive – it can be zero or it can be negative!

SO, WHEN **TIME** IS INCLUDED WE MUST MODIFY THE PYTHAGORAS THEOREM TO THE LORENTZ THEOREM ...

RECALL THE IDEA OF THE LORENTZ TRANSFORMATIONS ON PAGES 18 AND 19. WE NOW HAVE ...

$$ds^2 = -c^2dt^2 + dx^2 + dy^2 + dz^2$$

where $dt = t'-t$ is the difference in time between two events.

Our previous classification of geodesics into timelike, null and spacelike now corresponds to ds^2 being negative, zero and positive respectively.

Geodesics

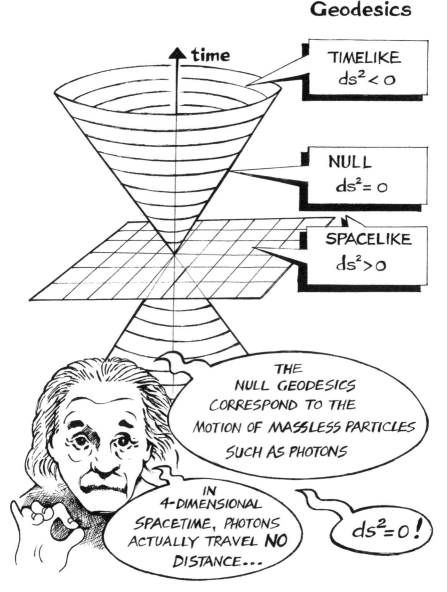

time

TIMELIKE
$ds^2 < 0$

NULL
$ds^2 = 0$

SPACELIKE
$ds^2 > 0$

THE NULL GEODESICS CORRESPOND TO THE MOTION OF MASSLESS PARTICLES SUCH AS PHOTONS

IN 4-DIMENSIONAL SPACETIME, PHOTONS ACTUALLY TRAVEL NO DISTANCE...

$ds^2 = 0$!

Though, of course, photons can travel great distances in SPACE.

The Dragon's Tail

We therefore see that a number of beautiful and radical extensions
of Newtonian gravity are implicit in our change from Newton's first law to
Einstein's modification. It only required us to change a couple of words.
Therein lies the amazing power and economy of GR which led the
famous Russian physicist **Lev Landau** (1908–68) to claim that a deep
awe and appreciation of GR are a prerequisite for being a theoretical
physicist.

If you think a little you may realize that there is a huge missing ingredient
which is fundamental to completing our aim of a consistent relativistic
replacement of Newton's theory of gravity.

The Missing Ingredient

That missing ingredient is contained in the question: "How does spacetime know how to curve to give the right geodesics to send the moon sailing in an ellipse around the earth?"

Since it is the earth's gravity that causes the moon to revolve around it, we know that mass must be doing the job of curving spacetime.

The Dragon Bites its Tail

So, to put it laconically, matter tells the geometry how to CURVE, while the geometry tells the matter how to MOVE.

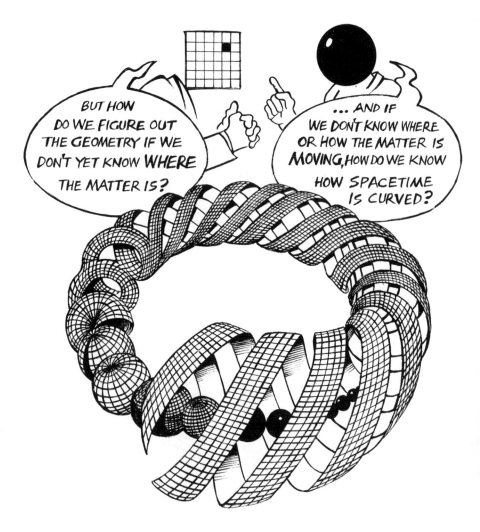

This chicken-and-egg situation is an intrinsic source of the complexity embedded in GR. We may think of a dragon which is biting its own tail.

Tensors

0-tensor is simply a single number: for example, the number "2"
1-tensor is a string of four numbers (in four spacetime dimensions).

So, for example, $A = (1 \quad 0 \quad -1 \quad 3.14)$ is a 1-tensor or simply just a "vector", which is also an arrow in spacetime.

Often we write A_i for the vector.

Here $i = 1, 2, 3$ or 4 so that $A_1 = 1$, $A_2 = 0$, etc. The electric and magnetic fields are described this way.

2-tensor is a matrix or block of 4 x 4 = 16 numbers which we can denote by B_{ij}. The two indices i and j tell us it is a block of numbers …

$$B_{ij} = \begin{pmatrix} 2 & 1 & 2.34 & 17 \\ -29 & 2 & 0 & 42 \\ 34 & -1.4 & 23 & 1000 \\ -1 & -1 & -1 & 0 \end{pmatrix}$$

Here the index i under B denotes the row number, while j denotes the column number. So, $B_{11} = 2$, $B_{12} = 1$, $B_{31} = 34$, and so on.

NOTE THAT THE ACTUAL NUMBERS IN THE BLOCK ARE NOT IMPORTANT…

…THEY COULD BE ANYTHING

3-tensor is a three-dimensional block of numbers which we can denote with three indices. For example, C_{ijk} where each of i, j, k can be any of 1, 2, 3 or 4.

This picture only shows some of the 4 x 4 x 4 = 64 numbers in this 3-tensor.

Einstein's Field Equations

Now that we have introduced tensors, which are the appropriate way to describe curvature of spaces, we can write down Einstein's field equations. But before this we need one more piece.

If we write $C_{ij} = B_{ij}$ then we mean that $C_{11} = B_{11}$, $C_{12} = B_{12}$, $C_{22} = B_{22}$, etc. … for all values of i and j.

Now we can write down Einstein's equations of General Relativity:

$$G_{ij} = 8\pi G T_{ij} + \Lambda g_{ij}$$

Or to write it out fully ...

$$\begin{pmatrix} G_{11} & G_{12} & G_{13} & G_{14} \\ G_{21} & G_{22} & G_{23} & G_{24} \\ G_{31} & G_{32} & G_{33} & G_{34} \\ G_{41} & G_{42} & G_{43} & G_{44} \end{pmatrix} = 8\pi G \begin{pmatrix} T_{11} & T_{12} & T_{13} & T_{14} \\ T_{21} & T_{22} & T_{23} & T_{24} \\ T_{31} & T_{32} & T_{33} & T_{34} \\ T_{41} & T_{42} & T_{43} & T_{44} \end{pmatrix}$$

$$+ \Lambda \begin{pmatrix} g_{11} & g_{12} & g_{13} & g_{14} \\ g_{21} & g_{22} & g_{23} & g_{24} \\ g_{31} & g_{32} & g_{33} & g_{34} \\ g_{41} & g_{42} & g_{43} & g_{44} \end{pmatrix}$$

Where G is Newton's gravitational constant, Pi is 3.14 ... and Λ is a constant known as the "cosmological constant". We will need it later. We see that Einstein's equations are actually 16 equations of the form ...

$$G_{11} = 8\pi G T_{11} + \Lambda g_{11} \qquad \text{and so on ...}$$

TENSORS ARE THEREFORE A VERY USEFUL AND COMPACT WAY TO WRITE THEM

In particular, if there is **no matter** at a particular point **(x,y,z,t)** – a vacuum – then T_{ij} **(x,y,z,t) = 0**.

From Einstein's equations this means that $G_{ij} = \Lambda g_{11}$ at the point **(x,y,z,t)**. But crucially, even if $\Lambda = 0$, this does NOT mean that the space is flat at the point **(x,y,z)**.

This is very important since from our own daily experience, the earth goes round the sun, even though the space between the sun and the earth is almost a perfect vacuum.

In general T_{ij} (x,y,z,t) is NOT zero, and we have to solve the 16 equations simultaneously – a very difficult task which we are still trying to do in general.

To proceed further, we must delve a little into the different types of curvature that a space can have – *intrinsic curvature and extrinsic curvature*.

Types of Curvature

Now that we have derived Einstein's equations, let's get a better handle on the different types of curvature we can expect to run into, as this will prove very useful later on. First, let's think about two-dimensional surfaces, such as the surface of the earth or a sheet of paper.

Euclid formulated the foundations of geometry.

In the end, Euclid had to take it as an assumption – an **axiom**. This is because it is NOT generally true.

In fact, it is true in general only if the space on which you draw the parallel lines is flat. Hence, Euclidean geometry is the study of flat-surface geometry!

Positive Curvature

To see that parallel lines can meet, we need to have a definition that is suitable for curved spaces. Now, each of the two parallel lines in Euclid's geometry are straight lines, so it seems obvious (after our modification of Newton's first law to Einstein's) that we should substitute "geodesic" instead of "straight line" in the general definition of parallel lines.

WE THEN SAY TWO GEODESICS ARE PARALLEL IF THEY ARE PARALLEL AT SOME POINT...

... THAT IS, THE ANGLES THEY MAKE ON INTERSECTION WITH A THIRD GEODESIC ARE THE SAME

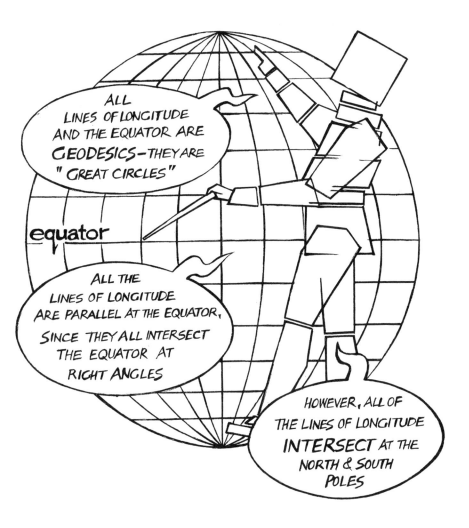

So parallel lines can meet! In this case, the space is said to have POSITIVE curvature.

Negative Curvature

It is also possible to construct spaces in which the parallel geodesics never intersect, but the distance between them increases the further along the geodesics you go.

Finally, we have the flat spaces of Euclid where parallel lines remain equidistant and never meet.

Another interesting way we can characterize these three different types of curvature (FLAT, POSITIVE, NEGATIVE) is by generalizing the idea of a triangle. Usually (i.e. in a flat space) a triangle has three sides made from straight lines. In a negatively or positively curved space, straight lines often don't exist!

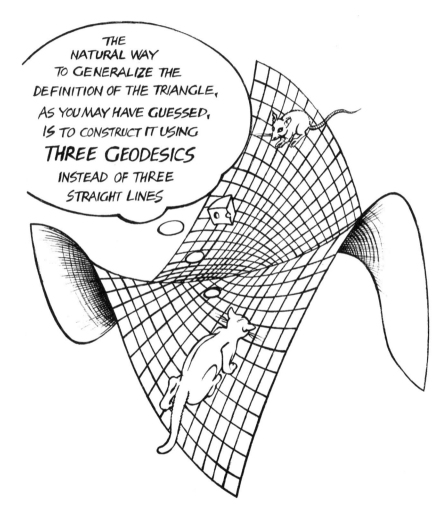

This definition automatically reduces to the usual one in a flat space, since the geodesics on a flat space are straight lines.

Triangles in Curved Space

Now we are in a position to ask questions about the properties of these generalized triangles. For example, what happens to fundamental schoolbook theorems, such as "the internal angles of any triangle add up to 180 degrees"?

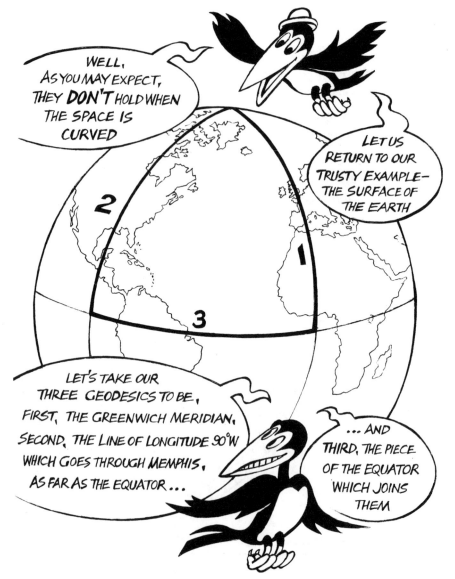

Positively Curved ...

This is a general characteristic of **positively** curved spaces – the sum of angles of triangles formed from their geodesics is GREATER than 180 degrees.

Intrinsic Curvature

We now have to address some interesting subtleties to do with formulating general relativity in three space dimensions and one time dimension. To appreciate why this is important, we must introduce two new aspects of curvature. We had classified the curvature of a space by whether parallel geodesics *intersect* or *diverge*, and then by the sum of the interior angles of geodesic triangles.

THIS CONTROLS WHAT WE WILL CALL INTRINSIC CURVATURE. BUT THERE IS ANOTHER TYPE OF CURVATURE

SUPPOSE I DRAW TWO PARALLEL GEODESICS—STRAIGHT LINES—DIAGONALLY ACROSS THE PAPER

OF COURSE THEY DO NOT INTERSECT...

Extrinsic Curvature

NOW, IF I WRAP THE SHEET OF PAPER ROUND ON ITSELF AND GLUE THE EDGES TO FORM A CYLINDER...

...THE STRAIGHT LINES I DREW ARE NO LONGER STRAIGHT, BUT THEY ARE STILL PARALLEL, AND THEY STILL DON'T INTERSECT!

But a cylinder is clearly not flat! How can this be?

Well, because parallel geodesics remain equidistant, we know that the cylinder, like the flat sheet of paper, is INTRINSICALLY FLAT. However, it is intuitively clear that in some way the cylinder really is curved. And at the same time, it is intuitively obvious that a flat piece of paper really is flat.

What is the main difference between the two cases?

So the difference obviously has to do with the way the cylinder looks as a whole. Or the way this two-dimensional space is placed – or embedded – in three dimensions.

This means we need another type of curvature, which is known as EXTRINSIC CURVATURE. But how do we quantify extrinsic curvature?

With a little thought, the correct solution may pop up.

Normal Vectors

Let's think of the flat sheet of paper again. Let us construct a line perpendicular to the sheet of paper, going through it at some point (x,y).

Then do the same with the cylinder. Now things are more interesting. The normal vectors lie on lines which emanate from the centre line of the cylinder.

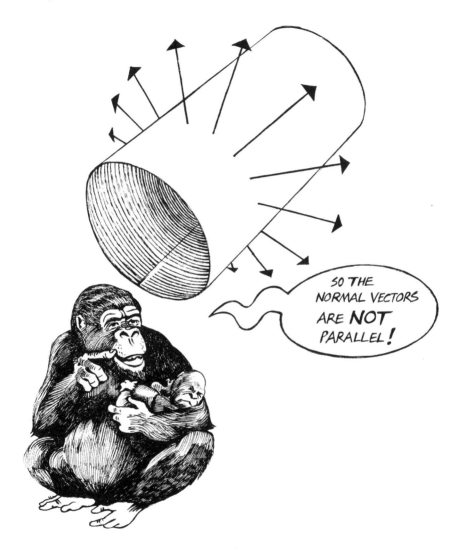

SO THE NORMAL VECTORS ARE **NOT** PARALLEL!

This is the key we have been looking for – the normal vectors are not all parallel in a space which has some extrinsic curvature.

Spatial Slices

The concepts of intrinsic and extrinsic curvatures are particularly useful for spacetime physics, because spacetime has three space dimensions and one time dimension.

IT IS OFTEN HELPFUL TO SLICE SPACETIME INTO 3-DIMENSIONAL SPATIAL SECTIONS WHICH, WHEN STACKED TOGETHER, FORM OUR 4-DIMENSIONAL SPACETIME

We can ask both about the intrinsic and extrinsic curvature of the three-dimensional spatial sections. Indeed, both are required to understand the curvature of four-dimensional spacetime. We could then rewrite Einstein's equations in terms of the intrinsic curvature and extrinsic curvature of the three-dimensional spatial slices.

Note that we have only considered spaces where the curvature is constant all over the spaces – a sphere, a cylinder, a flat sheet of paper. They are easy to visualize but extremely special. Most spaces will have curvature that varies over the space.

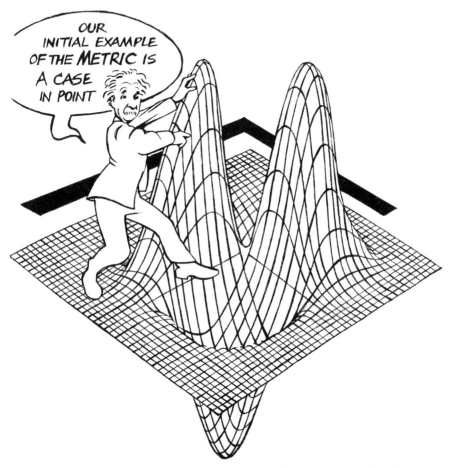

The curvature of a landscape is large in mountainous regions with steep valleys but small on the grass lands, which are nearly flat. Hence, while the earth is approximately a sphere (and the curvature is approximately constant), it has small variations in its curvature due to landscape variations.

The same is true of the curvature of spacetime in general relativity, as we will see.

Space and Time vs. Spacetime

What is going on? First, we emphasized how Einstein unified space and time into spacetime. Now we are talking of space and time again and have stated that we can rewrite Einstein's equations in terms of intrinsic and extrinsic curvatures. To make sense of this, let's return to Special Relativity.

Depending on how they move through spacetime, the slicing is relative and is not absolute!

In the case of a curved spacetime, we can imagine an infinite number of observers at each point of space who are moving differently in general.

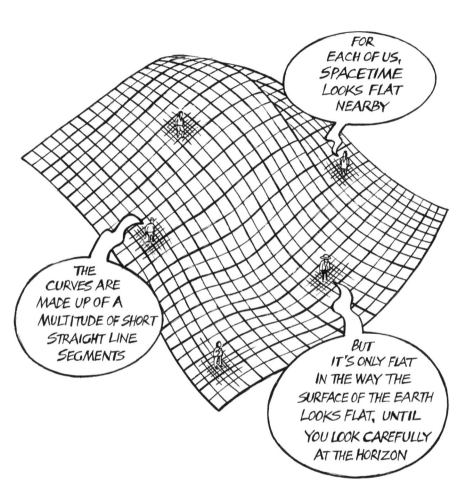

But when one considers all the observers, the spatial sections will generally be far from flat.

Splitting space and time is particularly important when we come actually to solve Einstein's equations and apply them to realistic situations, such as building models of the universe.

Testing GR in Nature

Einstein's equations satisfy the basic requests we had. But the final judge, after the conception of a new theory, is always Nature. What tests has general relativity been subjected to? What predictions have been verified? We have already discussed a known problem with Newtonian gravity – the perihelion shift of the planet Mercury.

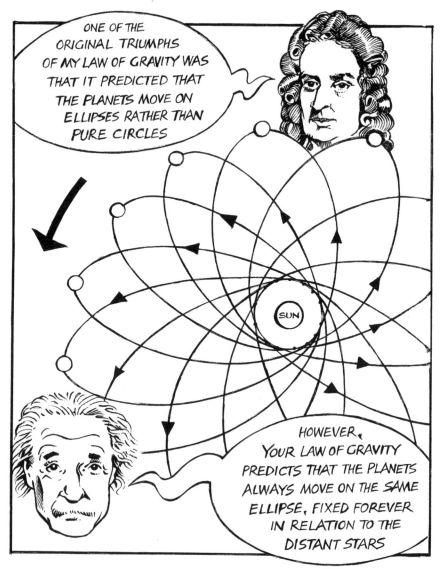

So it was a considerable quandary when observations of Mercury's orbit showed that the point closest to the sun – the perihelion – was in a slightly different place every orbit.

The Bending of Light

Predicting Mercury's perihelion was not sufficient to convince everyone about the reality and usefulness of GR. Einstein won the Nobel Prize in 1921 for the so-called photoelectric effect and contributions to theoretical physics, and not for GR.

In 1919, this was proven ...

This was famously tested by the astronomer **Sir Arthur Eddington** (1882–1944). His expedition sailed from England in March 1919 to Principe Island, off west-coast Africa, to study an eclipse of the sun.

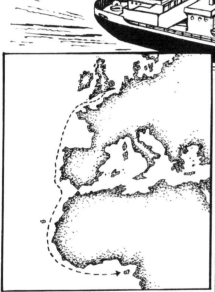

The eclipse was due to occur at 2 p.m. on 29 May, but there was a storm with heavy rain that morning. Eddington wrote: "The rain stopped about noon and about 1.30 … we began to get a glimpse of the sun. We had to carry out our photographs in faith …"

The Eclipse

"I did not see the eclipse, being too busy changing plates, except for one glance to make sure that it had begun and another half-way through to see how much cloud there was. We took sixteen photographs. They are all good of the sun, showing a very remarkable prominence; but the cloud has interfered with the star images. The last few photographs show a few images which I hope will give us what we need …" Eddington then later wrote …

Einstein's prediction was that starlight would be bent by the sun a factor of two greater than that predicted by Newton's theory of gravity. Eddington's observations gave convincing evidence of general relativity's validity. Eddington later composed the stanza ...

> *Oh leave the Wise our measure to collate,*
> *One thing at least is certain, light has weight.*
> *One thing is certain and the rest debate:*
> *Light rays, when near the Sun, do not go straight.*

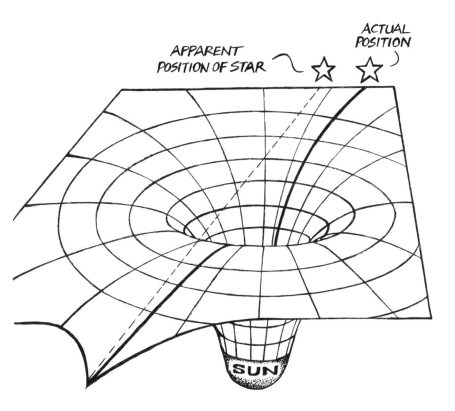

The Equivalence Principle Again

We have alluded to another prediction of GR – the equivalence of inertial and gravitational mass – expressed as the **equivalence principle**. This principle implies that all bodies will fall towards the earth with exactly the same acceleration, to an equal **non-gravitational** force acting on the body.

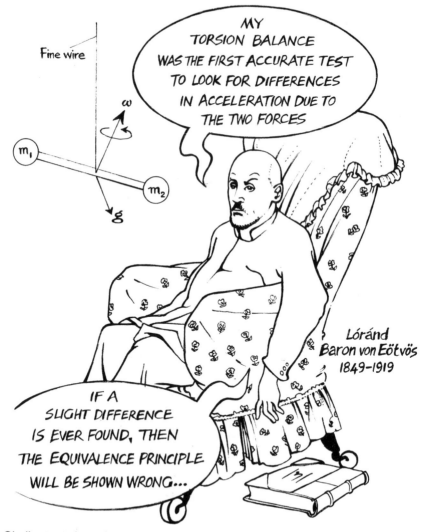

Similar tests have been regularly carried out, each with better accuracy, but so far no difference has ever been measured.

The Best-tested Theory

Today, Einstein's theory is – perhaps with the exception of quantum electrodynamics – the best-tested theory ever. However, there are good reasons to believe that GR must break down in certain circumstances.

There are further key predictions of GR that await full detection – **black holes** and **gravitational waves**.

Black Holes

Loosely speaking, Einstein's equations state that the more matter there is in a region, the more spacetime **curves** in that region. Hence, the more the matter is drawn into that region, and the harder it is for a body to escape.

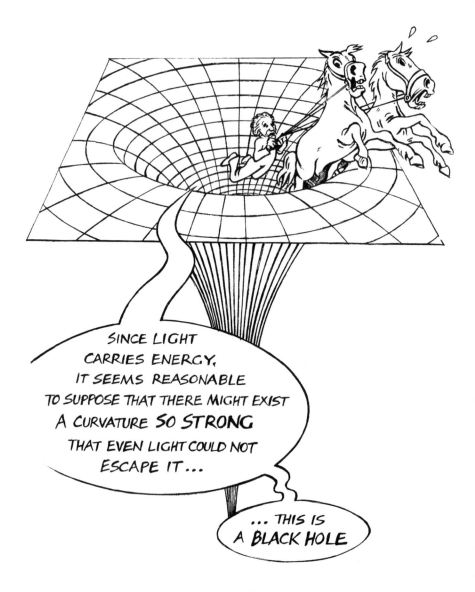

SINCE LIGHT CARRIES ENERGY, IT SEEMS REASONABLE TO SUPPOSE THAT THERE MIGHT EXIST A CURVATURE **SO STRONG** THAT EVEN LIGHT COULD NOT ESCAPE IT...

... THIS IS A **BLACK HOLE**

The first black-hole solution was found by the German mathematician **Karl Schwarzschild** (1873–1916) in 1916.

It is believed that fantastically massive black holes, with masses around one million times the mass of our sun, lie at the centres of many galaxies including our own.

Time-varying Acceleration

The other key prediction of GR is gravitational waves. Let us ask the question, "When is light emitted?" Light is classically described by oscillating electric and magnetic fields.

In short, the varying magnetic field will cause a varying electric field, and so on. This is an **electromagnetic wave**. Consider what happens when an electromagnetic wave passes through a radio antenna.

So, electromagnetic waves are emitted when electric charges are accelerated – as in a radio antenna.

Shaking a Mass

What happens when we shake a mass – for instance, a star – back and forth? "Shake a mass" means literally to make an object with some mass move back and forth.

It turns out that GR predicts that a mass undergoing time-varying acceleration should emit gravitational waves. Now, what are these waves?

The Rubber-Sheet Analogy

Another simpler way to think of gravitational waves is to compare them to a stretched sheet of rubber.

In the same way, gravitational waves will spread out in all directions around the shaken mass.

Gravity's Weakness

However, because Newton's constant G is so tiny, these gravity waves are incredibly weak – if indeed they exist at all.

Well, light carries energy – e.g., one gets sunburned on the beach!

Hence, gravitational waves should also carry energy.

We may therefore hope to see a body either losing or gaining energy due to **emission** or **absorption** of gravitational waves.

Stargazing

The best evidence we currently have of gravitational waves comes from observations of a now-famous pair of stars – a binary system called PSR 1913 + 16 – rapidly orbiting around each other. Studied for over 25 years, accurate observations have shown that – like the perihelion of Mercury – the period of the orbit is not constant.

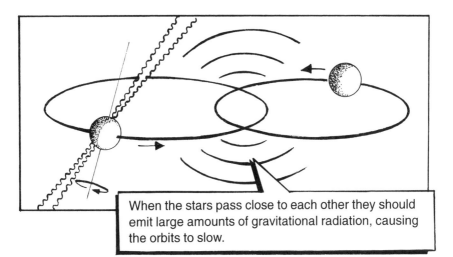

When the stars pass close to each other they should emit large amounts of gravitational radiation, causing the orbits to slow.

For their work, Hulse and Taylor won the 1993 Nobel prize for physics since it provides an exquisite test of GR.

Still, the slow-down in the orbiting of the binary system might plausibly be due to something else – though this is admittedly unlikely. It is generally agreed – in accord with the scientific spirit – that direct detection of gravity waves is needed finally to prove the existence of gravitational waves.

SO FAR, EXPERIMENTS HAVE NOT DETECTED THEM

BUT THAT IS CONSISTENT WITH OUR EXPECTATIONS THAT THEY ARE SO WEAK

The first decade of the 21st century should certainly yield a positive detection – if they exist – by directly using the fact that gravitational waves would stretch and compress spacetime.

Interferometric Observation

How are we to detect gravitational waves **directly**? Imagine that you hope to see the stretching of spacetime directly using a metre-stick.

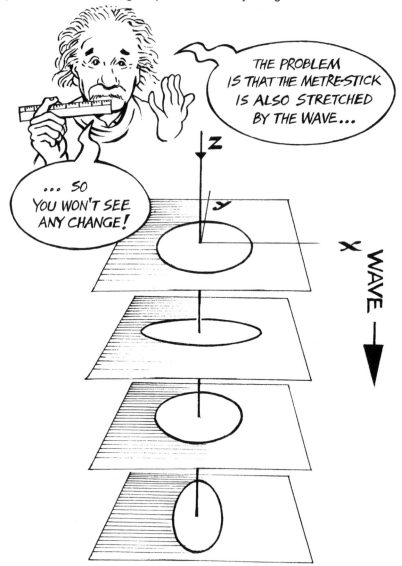

A gravitational wave travelling along the z direction will deform a circle into an ellipse, first along the x axis and then along the y axis, and so on, until it has completely passed by.

A new generation of gravitational-wave telescopes is just about ready (as of 2002) to run and give results soon. In the USA, there is LIGO (Laser Interferometric Gravitational Observatory). The UK has GEO600 (a joint project with Germany). There is the French-Italian VIRGO. Japan has TAMA. All of these systems are very expensive and based on laser interferometers.

LIGO Hanford Observatory, Richland, Wa.
(A second giant interferometer is in Livingston, La.)

How it Works

An interferometer is a rather simple device. It consists of two arms at right angles.

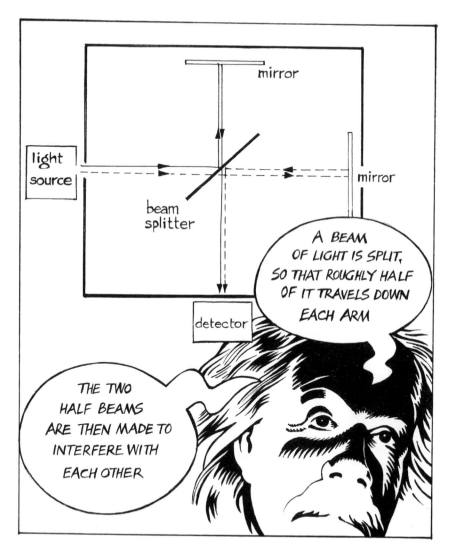

This results in a characteristic light-dark-light interference pattern – which shows that light has wavelike properties.

The dark patches occur where the light from the two arms was completely out of phase. A peak in one arm meets a trough from the other. The bright patches occur where the light was completely in phase – two troughs or two peaks.

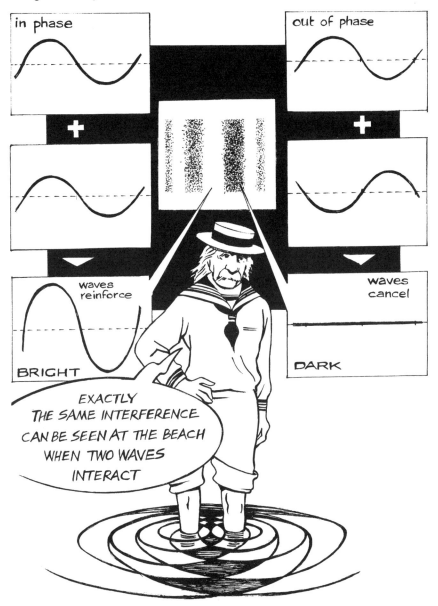

Interference Patterns

What is the idea behind using an interferometer? If a gravitational wave passes by and stretches one of the arms of the interferometer, then the path that light must travel along that arm before being reflected by the mirror is **increased**.

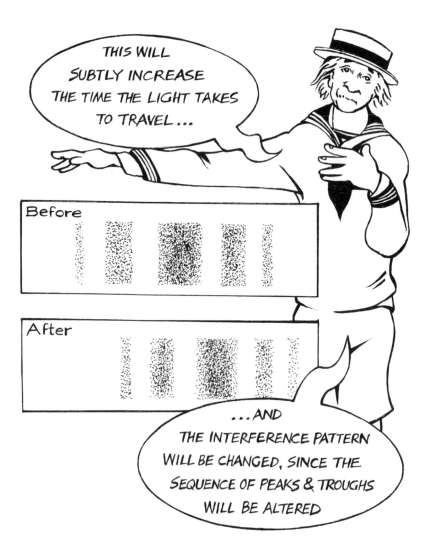

Black holes and gravitational waves are two exciting predictions. They typically occur on relatively small scales. Next, we consider what happens on a grand scale when we consider the universe as a single object. We try to understand *where* it came from and *where* it is going, using Einstein's equations.

Sizing Up the Universe

If we look out at the universe around us, we first see the planets in our solar system. Beyond them, we see the stars and gas clouds of our galaxy which is several thousand light-years across. Remember, a light-year is the enormous distance travelled by light in a year!

Outside our own galaxy, we see about 100 billion galaxies. A natural question would be, "How are these 100 billion galaxies distributed around us?"

For example, are the galaxies clumped in one direction? It turns out remarkably that they are in fact uniformly spread around us on the sky.

But first let us focus on a slightly different issue.

The Copernican Principle

Since the number of galaxies is roughly the same in all directions, this can mean either of two things.

The history of cosmology – and physics in general – has been rather at odds with the Christian church.

The idea that we are "near" the centre of the universe has proven very unpopular.

This is partly because of the non-religious bias, and partly because the resulting cosmological models are significantly simpler.

Simplifying the Field Equations

"Simpler models" are definitely a bonus for cosmologists. It should be emphasized that Einstein's field equations are tremendously complex and have remained generally unsolved since their discovery.

"FLRW"

Soon after GR was proposed, cosmologies were found that obeyed the Copernican principle. These "simplifications" were the work of four scientists, known by the acronym "FLRW" – the Russian **Alexander Friedmann** (1888–1925), the Belgian priest **Georges Lemaître** (1894–1966), the North-American **H. P. Robertson** and the English mathematician **Arthur G. Walker.**

WE DID THIS BY DESCRIBING "CASES OF HIGH SYMMETRY" – THAT IS, EXPANDING UNIVERSES

Georges Lemaître

WE "FOUND" COSMOLOGIES BY FINDING SIMPLE SOLUTIONS TO THE EQUATIONS WHICH DESCRIBE EXPANDING UNIVERSES

I'M NOT CONVINCED BY YOUR "EXPANDING UNIVERSES"...

Static or Expanding Universes?

FLRW generalized the work begun by Einstein. But, the problem was, Einstein found that he could not have a static universe unless he introduced a "cosmological constant" – a repulsive force – to balance the attractive force of gravity.

FLRW models, as they are known, form the backbone of cosmology. Everything you may hear regarding cosmology in popular literature is almost certainly based on these models.

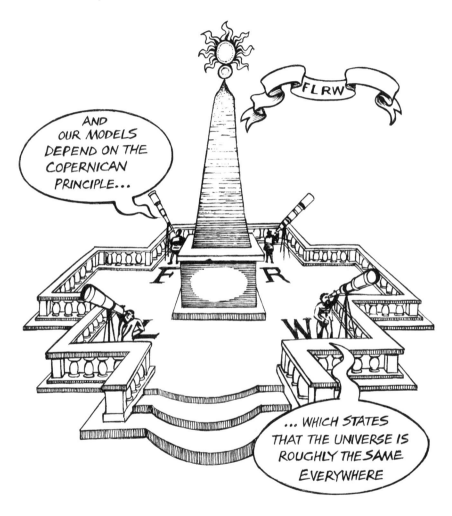

The Copernican principle has remained remarkably difficult to prove – or disprove – though significant progress towards this goal should be made in the next 15 years.

The Fate of the Universe

One of the elegant features of the FLRW models is that there are basically only **three** of them. In other words, there are only three different types of FLRW solutions to Einstein's field equations, each classified by their **curvature** – the three options being positive, negative and flat. All the models begin with a "Big Bang" – a term coined disparagingly by the cosmologist **Sir Fred Hoyle** (1915–2001).

But the subsequent evolution of each FLRW model is radically different – and so is the fate of the universe!

The Critical Density: First Model

In these models, there is a critical density of about 10^{-29} grams per centimetre cubed. The "critical density" refers to the density of all types of matter and radiation added together, i.e. hydrogen, light, dark matter, cosmological constant – everything. Above this density, the universe is finite, its spaces are three-dimensional spheres, or in other words, **positively curved**.

Second Model

Below this critical density, the universe has more kinetic energy, roughly speaking, than the gravitational force can rein in.

In this case, the universe is infinite both in space and time (since it lasts for ever) and the spaces are *negatively curved*.

Third Model

At exactly the critical density, the three-dimensional space is precisely flat – the analogue of a two-dimensional sheet of paper.

$k = O$

radius of universe

time

ALTHOUGH THE CRITICAL DENSITY OF 10^{-29} IS INCREDIBLY SMALL, OUR UNIVERSE IS VERY CLOSE TO IT ON AVERAGE

WHICH SIDE WE ARE ON EXACTLY—ABOVE, BELOW, OR AT CRITICAL DENSITY—IS STILL UNKNOWN...

Explaining Redshift

In 1929, the astronomer **Edwin Hubble** (1889–1953) discovered the expansion of the universe experimentally by seeing that the dimmer a galaxy was, the more its light was "redshifted" to longer wavelengths.

Edwin Hubble

THE SIMPLEST EXPLANATION IS THAT THE DIMMER GALAXY, FURTHER FROM US, IS **MOVING AWAY FROM US** FASTER THAN NEARER GALAXIES

In visible light, higher frequencies appear blue, while lower frequencies appear red. If an object emits light while moving rapidly away from an observer, then that light will appear redder than it did when the object was not moving away from the observer.

Einstein's Static Universe

Hubble's observation was a crucial turning-point. Einstein realized he had missed the chance to predict that the universe was expanding by assuming that the universe HAD to be static.

143

The Accelerating Universe

Since the cosmological constant can be repulsive, it can act like matter with negative pressure and would actively push galaxies away from each other. Conversely, the further away galaxies get from each other, the weaker the attraction of gravity they feel for each other.

But the repulsive cosmological constant does not get weaker with increasing distance.

In other words, they accelerate away from each other – in cosmic jargon, the universe begins to **accelerate**.

Endless Expansion

Acceleration is a profound effect since it can alter the future destiny of the universe. If the universe begins to accelerate due to Lambda, it will very likely – if general relativity is correct and barring "strange matter" – accelerate for ever.

Recent observations of distant supernovae (giant cosmic explosions) appear to show that these are dimmer than one would expect in a universe which had not been accelerating. If the universe has been accelerating, then objects at a fixed redshift are further away than in a non-accelerating universe – and hence they appear dimmer.

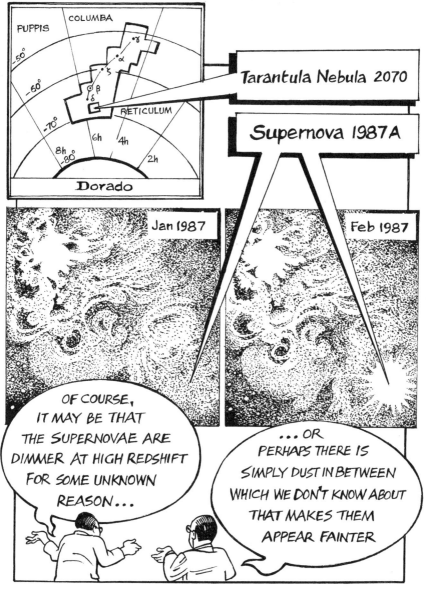

Negative Pressure

Nevertheless, evidence from supernovae combined with observations of the Cosmic Microwave Background (CMB) – as we'll see later – makes it seem pretty certain that there is a large portion of "matter" out there with negative pressure, probably making up at least 60% of the universe's total energy density. Note that this energy is *not* the same as Dirac's anti-matter.

This large portion of negative energy is not the only cosmologically important "matter mystery" out there. It has been known now for several decades that galaxies appear to rotate *incorrectly*.

Dark Matter

One can see this problem by thinking of gravity as the force that keeps the stars moving in a circular orbit, like a rope holding a whirling stone in a circle.

However, if we estimate the mass of the galaxy from the amount of matter we can see, there isn't enough to keep the stars in their circular orbits at the speeds we observe them moving at. This is the "rotation curve problem".

In fact, it appears that at least 25% of the energy of the universe is made from this dark matter, which we have never directly detected!

Beyond General Relativity

A natural question might be – isn't this dark matter just like the perihelion shift of Mercury, the effect that couldn't be explained by Newton's theory of gravity?

Perhaps there are candidates for this that might come from particle physics – the physics of the very small.

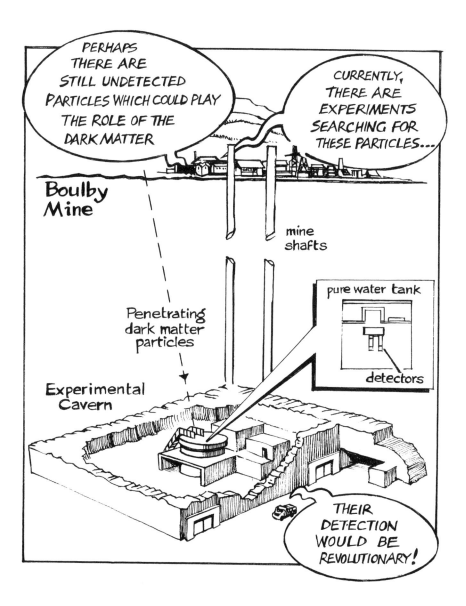

The Cosmic Microwave Background

In the 1960s, two physicists at the Bell Telephone laboratories in New Jersey, USA, Arno Penzias and Robert Wilson, found a strange static in every direction when they looked at the wavelength of microwaves.

It took three decades of hard searching to find any variations in the temperature across the whole sky.

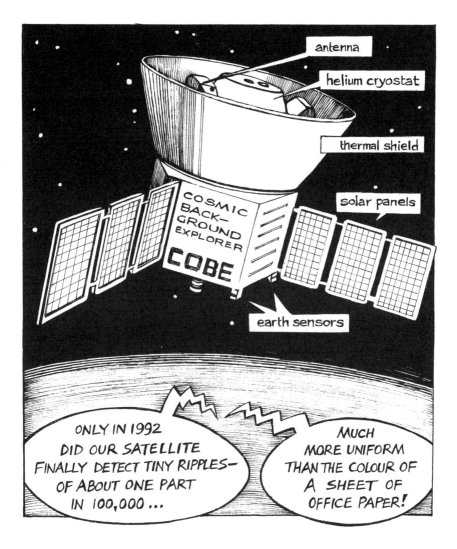

Further Satellite Probes

The year 2001 saw the launch of another CMB satellite – the Microwave Anisotropy Probe (MAP) – much higher in precision than COBE.

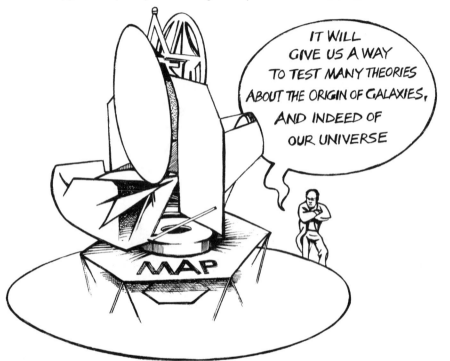

IT WILL GIVE US A WAY TO TEST MANY THEORIES ABOUT THE ORIGIN OF GALAXIES, AND INDEED OF OUR UNIVERSE

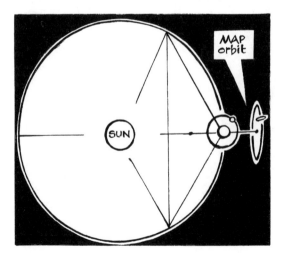

In 2007 (or shortly thereafter) another CMB satellite, PLANCK, will be launched with even higher precision instruments. These two experiments will shed enormous light on our understanding of the universe.

The Homogeneity Mystery

Why do we care about the CMB? Well, the fact that cosmic radiation is so incredibly uniform in temperature is a mystery. Imagine taking a huge bag of, say, a million coins and emptying it over the carpet and then finding that all of them landed heads up, except for only ten tails.

The average number of galaxies in every direction seems to be the same. Why is this and how did it happen? This is the "homogeneity" problem.

The "Goldilocks" Expansion Rate

But the paradoxes get worse, even if we choose a perfectly uniform distribution of galaxies. FLRW cosmologies listed three basic types: those with three-dimensional spatial slices which are, respectively, negatively curved, flat and positively curved. Which one our universe corresponds to depends on its average density.

But here is the problem. If the universe is far from flat, it would either have recollapsed before now – our universe has lasted at least 10 billion years – or no galaxies, stars and planets would have been able to form due to the rapid expansion of the universe.

The Flatness Problem

There would be no problem except for the flat-case universe being unstable. It is like balancing a pencil on its tip. If you push it ever so slightly in any direction then the pencil will fall over.

This is known as the "flatness problem" and, with the earlier homogeneity problem, these are two of the enduring mysteries of Einstein gravity and cosmology today.

The Inflation Phase

The flatness and homogeneity problems have been known for several decades now. In 1980, Alan Guth at the Massachusetts Institute of Technology (MIT) in the USA had an idea which (although independently and partially discussed earlier by several researchers) became a major contribution to cosmology.

Using Einstein's Constant

Remember that Einstein's "biggest mistake" was to introduce the repulsive cosmological constant to keep the universe static. Guth, in contrast, proposed to use this repulsive force to make the universe accelerate very rapidly – much more rapidly than we think it is accelerating today.

In this sense, inflation makes our pencil want to stand up on its tip again. If the universe inflated enough, it would then be no surprise that we see the universe as we do.

Guth also proposed a new way to get this acceleration, based on a new type of matter, rather than using the cosmological constant. Guth's new "scalar field" matter has not so far been detected – and despite being predicted by our near-latest theories of particle physics. However, short of throwing away Einstein's equations, we have no other widely accepted explanation for the observed characteristics of the universe other than inflation.

With the CMB, MAP and PLANCK satellites, we hope to test inflation.

Singularity Theorems

As we have mentioned, black-hole solutions were known to Einstein and other relativists.

Hawking showed that any realistic cosmology, which obeys the Einstein equations, MUST have had a point – a FINITE amount of time in its past – at which the density and the curvature of the universe were infinite. This is the Big Bang and the corresponding theorems (for the black hole and universe) are known as the **singularity theorems**.

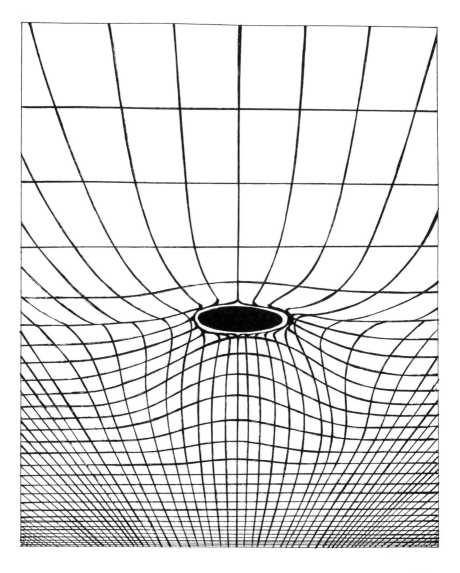

The Result of Singularity Theorems

The singularity theorems guarantee us that even if gravity and the density of the universe are, respectively, very weak and very small now, they must have been infinitely large in the past, at a time conservatively estimated to be between 10 and 18 billion years ago.

This collapse ends at the Big Bang when that inter-particle distance becomes zero.

Why is this result remarkable? It is amazing for a number of reasons. First, when the curvature is infinite we are unable to use GR theory! It stops making predictions.

The speed of light, c, and Newton's constant, G, are the only constants of the theory. But Planck's constant, h, appears nowhere.

The Invalidation of Einstein's Equations

Why is Planck's constant important? When the average inter-particle distance becomes about the same size as an atom, then classical physics fails and quantum effects start to become important. As the density of the universe steeply rises, most cosmologists believe there must come a point at which Einstein's equations fail and are simply wrong, since they don't include any quantum effects.

We therefore need to extend Einstein's equations to include quantum effects. The most famous physicists of the 20th century, including Einstein himself, have tried and failed to unify GR and quantum theory.

Extra Dimensions

Einstein hoped to find a unified theory – like that of electromagnetism – describing all of the forces of nature, based on geometry in the same way as GR is.

Two physicists, Theodor Kaluza and Oskar Klein, took a step towards Einstein's dream when they decided to look at a five-dimensional world.

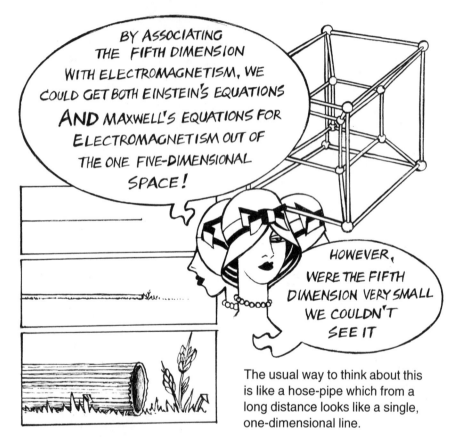

BY ASSOCIATING THE FIFTH DIMENSION WITH ELECTROMAGNETISM, WE COULD GET BOTH EINSTEIN'S EQUATIONS AND MAXWELL'S EQUATIONS FOR ELECTROMAGNETISM OUT OF THE ONE FIVE-DIMENSIONAL SPACE!

HOWEVER, WERE THE FIFTH DIMENSION VERY SMALL WE COULDN'T SEE IT

The usual way to think about this is like a hose-pipe which from a long distance looks like a single, one-dimensional line.

While this addition was a radical point of view, we now know that there are at least four forces of nature – gravity, electromagnetism, the weak force and the strong force. Can we view them *geometrically*?

Superstring Theory

A way to include all the forces geometrically is through **superstring theory**. This approach to quantum gravity is extremely fashionable currently. It was originally formulated as a way of treating the strong force which holds the neutrons and protons together in atomic nuclei.

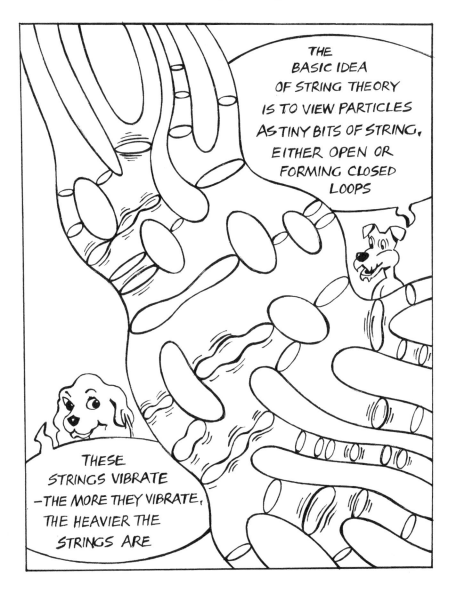

Extending Einstein's Dream

What is nice is that to make the string-theory work, the strings can vibrate only at certain specific frequencies – much like a guitar string. One finds that gravity is automatically included for free!

172

Adding More Dimensions

However, there is one very strange prediction of string theory. It predicts *extra dimensions*. In fact, it predicts that we live not in five, but rather in ten dimensions! Obviously we don't live in ten large dimensions.

Whether these extra dimensions exist or not will probably take a long time to test, since they would only be visible at very high energies. But the theory is certainly very elegant, in a way that would probably have pleased Einstein. It is certainly based on the fundamental ideas of relativity.

Suggested Further Reading

Introductory Level Books
These books are at an introductory level and should be accessible to a wide audience.

Einstein
Introducing Einstein, Schwartz and McGuinness (Icon Books, 1999).

General Relativity
The Meaning of Relativity, Albert Einstein (Princeton University Press, 1992)
Flat and Curved Spacetimes, George Ellis and Ruth Williams (Oxford University Press, 2000).

Cosmology
There is a huge choice of popular books in the general area of cosmology. Some recommended titles include:

Cosmology: A Very Short Introduction, Peter Coles (Oxford Paperbacks, 2001).
Just Six Numbers, Martin Rees (Orion fiction, 2001).
The Big Bang, Joseph Silk (W.H. Freeman & Company, 2001).
Between Inner Space and Outer Space, John Barrow (Oxford Paperbacks, 2000).

Inflation
The Inflationary Universe, Alan Guth (Vintage, 1998).

Quantum Mechanics
Introducing Quantum Theory, McEvoy and Zarate (Icon Books).
In Search of Schrödinger's Cat, John Gribbin (Corgi, 1985).

Quantum Gravity
Dreams of a Final Theory, Steven Weinberg (Vintage, 1993).
A Brief History of Time, Stephen Hawking (Bantam, 1995).
Introducing Hawking, McEvoy and Zarate (Icon Books, 1999).

Advanced Level

These require either more mathematical or physics background:

Subtle is the Lord, Abraham Pais (Oxford Paperbacks, 1984) – a famous and very insightful history of Einstein's life, giving wonderful insights into the development of relativity.

Introducing Einstein's Relativity, Ray d'Inverno (Clarendon Press, 1992). This is a good introduction to the mathematics and physics of GR. A standard advanced undergraduate/post-graduate text.

The classic advanced texts in GR are (notice how the first three all appeared in 1973):

The Large Scale Structure of Spacetime, Stephen Hawking and George Ellis (CUP, 1973).
Gravitation, Misner, Thorne and Wheeler (W.H. Freeman, 1973).
Gravitation and Cosmology, Steven Weinberg (John Wiley & Sons, 1972).
General Relativity, Robert Wald (Chicago University Press, 1984).

About the Author and Artist

Bruce Bassett is currently senior lecturer at the Institute of Cosmology and Gravitation of Portsmouth University where he studies (mostly unsuccessfully) the mysteries of the early universe. Previously he conducted research in the physics department of the University of Oxford and at the International School for Advanced Studies in Trieste. Earlier still he owned a web-development company in Cape Town and studied art. Bruce has over 30 published research articles, several friendly but rebellious PhD and ex-PhD students and lots of very smart collaborators, all of which allow him to travel the world frequently.

Ralph Edney trained as a mathematician, and has worked as a teacher, journalist and political cartoonist. He is the author of two graphic novels, and the illustrator of *Introducing Philosophy* and *Introducing Fractal Geometry*. He is also a cricket fanatic.

Acknowledgements

Bruce Bassett would particularly like to thank the staff at Icon Books and Peter Coles for getting him involved in this project. He also thanks Mike Bassett, Josh Bryer, George Ellis, David Kaiser, Philippos Papadopoulos, David Parkinson, Fabrizio Tamburini, Ran van der Merwe and Fermin Viniegra for creative suggestions and enlightening comments on the manuscript. BB gratefully acknowledges George Ellis for teaching him so much of the relativity he knows.

Index